ON THE PARROTS OF THE MALAYAN REGION, WITH REMARKS ON THEIR HABITS, DISTRIBUTION, AND AFFINITIES, AND THE DESCRIPTIONS OF TWO NEW SPECIES (1864)

BY

ALFRED RUSSEL WALLACE

British Library Cataloguing-in-Publication Data
A catalogue record for this book is available from the
British Library

Alfred Russel Wallace

Alfred Russel Wallace was born on 8th January 1823 in the village of Llanbadoc, in Monmouthshire, Wales.

At the age of five, Wallace's family moved to Hertford where he later enrolled at Hertford Grammar School. He was educated there until financial difficulties forced his family to withdraw him in 1836. He then boarded with his older brother John before becoming an apprentice to his eldest brother, William, a surveyor. He worked for William for six years until the business declined due to difficult economic conditions.

After a brief period of unemployment, he was hired as a master at the Collegiate School in Leicester to teach drawing, map-making, and surveying. During this time he met the entomologist Henry Bates who inspired Wallace to begin collecting insects. He and bates continued exchanging letters after Wallace left teaching to pursue his surveying career. They corresponded on prominent works of the time such as Charles Darwin's *The Voyage of the Beagle* (1839) and Robert Chamber's *Vestiges of the Natural History of Creation* (1844).

Wallace was inspired by the travelling naturalists of the day and decided to begin his exploration career collecting specimens in the Amazon rainforest. He explored the Rio Negra for four years, making notes on the peoples and languages he encountered as well as the geography, flora, and fauna. On his return voyage his ship, Helen, caught fire and he and the crew were stranded for ten days before being picked up by the Jordeson, a brig travelling from Cuba to London. All of his specimens aboard Helen had been lost.

After a brief stay in England he embarked on a journey to the Malay Archipelago (now Singapore, Malaysia, and Indonesia). During this eight year period he collected more than 126,000 specimens, several thousand of which represented new species to science. While travelling, Wallace refined his thoughts about evolution and in 1858 he outlined his theory of natural selection in an article he sent to Charles Darwin. This was published in the same year along with Darwin's own theory. Wallace eventually published an account of his travels *The Malay Archipelago* in 1869, and it became one of the most popular books of scientific exploration in the 19th century.

Upon his return to England, in 1862, Wallace became a staunch defender of Darwin's landmark work *On the Origin of Species* (1859). He wrote responses to those critical of the theory of natural selection, including 'Remarks on the Rev. S. Haughton's Paper on the Bee's Cell, And on the Origin of Species' (1863) and 'Creation by Law' (1867). The former of these was particularly pleasing to Darwin. Wallace also published important papers such as 'The Origin of Human Races and the Antiquity of Man Deduced from the Theory of 'Natural Selection" (1864) and books, including the much cited *Darwinism* (1889).

Wallace made a huge contribution to the natural sciences and he will continue to be remembered as one of the key figures in the development of evolutionary theory.

Wallace died on 7th November 1913 at the age of 90. He is buried in a small cemetery at Broadstone, Dorset, England.

ON THE PARROTS OF THE MALAYAN REGION, WITH REMARKS ON THEIR HABITS, DISTRIBUTION, AND AFFINITIES, AND THE DESCRIPTIONS OF TWO NEW SPECIES (1864)

The Psittaci or Parrots are an extensive and very isolated group of birds ranging over the tropics of the whole world, but, with the exception of those lands of anomalies, Australia and New Zealand, rarely found in the temperate and cooler regions. As nearly as I can estimate, the number of species of these birds known at present amounts to 365, grouped in about thirty-six genera and five families. The manner, however, in which these species and group are distributed over the globe is very remarkable. Taking the zoological regions established by Dr. Sclater, we find the following approximate numbers--

Regions.	Species.	Genera.	Families.	
Palæarctic....	0	0	0	14 genera, all of one family,
Nearctic	1	1	1	
Neotropical ..	150	9	1	
Ethiopian	25	4	1	
Indian	25	3	1	
Australian....	165	25	5	25 genera of 5 families,—

showing a remarkable poverty in the Indian and Ethiopian regions, both in species and groups; abundance of species in the Neotropical (S. American) region, with comparatively few genera; while the Australian region not only contains more species than the American, but possesses nearly three times as many genera, and nearly twice as many as all the other regions combined. More remarkable still, the whole of the Indian, Ethiopian, and American Parrots belong to one great family group, the *Psittacidæ*, indicating a general uniformity of organization; while those of the Australian region mostly belong to three other distinct families--the Brushtongues or Lories (*Trichoglossidæ*), the Cockatoos (*Plyctolophidæ*), and the Broadtails or Ground-Parrots (*Platycercidæ*)--together with a few of the *Psittacidæ*, which last, however, are confined to the Malayan portion of the region.

These facts are of the highest interest in their bearing on the probable origin of the whole Psittacine group; for it is natural to suppose that in that portion of the earth's surface where the species are now most numerous, the forms most varied, where the most singular modifications of structure occur, and where both the highest and the lowest developments of the group are to be found, would be its true metropolis and original birthplace. I believe, therefore, that the Parrot type originated in the Australian region--a region now consisting almost entirely of broken land and scattered islands, but which, there is every reason to think, was once a continental area.

Confining our attention now to the Australian region only, we may divide it into three subregions--Australia, the Pacific Islands, and the Austro-Malayan group--each of which has a distinctive character. The *Platycerci* and the Cockatoos are more particularly the features of Australia and Tasmania, which have also a few *Trichoglossi*, but no *Psittacidæ*. The *Coriphili*, Nestors, and *Strigopidæ* are confined to the Pacific Islands, which have also *Platycerci*, but no Cockatoos. The crimson Lories are entirely restricted to the Malayan district, which has also abundance of Cockatoos, but few *Platycerci*, and several peculiar genera of *Psittacidæ*. Thus four out of the five families into which the order is divided are found in the Austro-Malayan district; they all extend into every part of it, and they are all represented by abundance of species, and three of them by numerous peculiar genera and even subfamilies.

The Australian subregion possesses three of the families only, and has a smaller number both of genera and species; but it has a large proportion of peculiar genera, and is preeminent in its numerous forms of *Platycercidæ* and *Plyctolophidæ*. Only three families and four genera extend to the Pacific Islands, but three of the genera are quite peculiar to them.

The following table shows these proportionate numbers

at one view:--

	Families.	Genera.	Species.
Austro-Malayan Islands	4	16	86
Australia	3	10	60
Pacific Islands	3	4	30

We thus see that the preeminence both in species, genera, and families is with the Austro-Malayan region, and we have therefore an *à priori* reason for considering it to be the most ancient, and to suspect that it may contain within its limits a portion of the country to which the Psittacine order was at one period restricted. Let us therefore examine its productions a little more carefully.

Undoubtedly the most highly organized form of Parrot is the Trichoglossine or Brushtongue family, in which the whole structure is modified to enable these birds to derive a considerable portion of their subsistence from the nectar of flowers. The bill is unusually small, elongated, and compressed, so that it may readily enter the corolla; the tongue is large, long, and very extensible, and can be thrust down to the very bottom of the nectary; and the papillæ of the terminal portion of the upper surface are developed into erectile fibres, forming a double brush, which rapidly gathers up all the honeyed secretions of the blossom. In correlation with this structure the species are mostly of small size, of graceful forms, and have powerfully grasping feet--qualities that enable them to climb actively among the twigs and branches, and to cling in any position to the waving sprays of blossom. They have also elongated wings and a powerful flight, which give them the means to traverse the whole area of their range, and discover at the right moment the flowering trees which are so attractive to them, the period of blossoming in tropical regions being very limited for each species. These extremely interesting birds are spread over the whole Australian region, while not one of them has been found beyond its limits; but it is in the Austro-Malayan district only that they are very abundant and of the most varied forms; for it is here that four out of the six genera are exclusively found. Three of these genera form a natural subdivision of the family, which may be called the *Loriinæ*, and comprise all those beautiful birds the ground-colour of whose plumage is vivid crimson, and which are commonly known as Lories. Of these the genus *Eos* may be considered the most typical, since the species have completely lost the green colouring which is so characteristic of Parrots generally, and by their activity, elegance, and more powerful flight show that they are the most highly developed in the Trichoglossine series. The vivid red colour which is so characteristic of the Lories here reaches its maximum in such species as *E. rubra* and *E. cardinalis*. This fine group, consisting of three genera and at least eighteen distinct species, has a singularly restricted range, being confined to an elongated tract comprising New Guinea and the islands east and west of it, from the Moluccas to the Solomon Islands. If we look at this area marked out upon a map[1], we must be at once impressed with the idea that we have here roughly indicated the much greater extent at a recent period of the large island of New Guinea, the north-western portion of which seems even now to be undergoing a still further segmentation. This idea receives confirmation from the fact that almost every bird found in this area has its closest allies in New Guinea; and as we approach the central mass, the variety of forms becomes greater. The genera *Monarcha* and *Mimeta* here have their maximum development; and *Tanysiptera*, as abnormal among Kingfishers as Lories are among Parrots, has almost exactly the same limits of distribution as they have. Within these limits are found some of the most curious forms of Parrots, the giant black Cockatoo (*Microglossus*) and the dwarf of the whole order (*Nasiterna*), the bare-headed *Dasyptilus* and the elegant little *Charmosyna*. *Tanygnathus* has the range of *Eos*, but extends north-westwards to Celebes and the Philippines; *Geoffroyus* south-westwards to Timor and Flores; while the remarkable genus *Eclectus* has exactly the same range as the Lories, whose attire it seems to mimic, for in it alone are found Parrots whose only colours are red with a portion of blue and black. This coincidence of the range of the red *Psittacidæ* with that of the red *Trichoglossidæ* is a very curious fact, and clearly intimates that these gay tints are not mere sports of nature, or designed for the delectation of man, but have a close connexion with the life-history of the creatures which they adorn, and probably subserve an important though hidden object in the economy of these groups.

The whole number of Parrots known to inhabit this Loriine region is fifty-four, belonging to no less than fifteen genera, of which eight genera are altogether peculiar to it. This is very remarkable when we compare it with Australia, which, though many times more extensive and also exceptionally rich in this family, possesses, with a rather larger number of species, about ten genera, of which six or seven only are peculiar to it. But Australia has been comparatively well explored in every part of its great extent, and, though a few more species may be discovered, we cannot expect it to produce any new forms of Parrots. New Guinea, on the other hand, the centre and primary mass to which the surrounding islands are but satellites, is a *terra incognita.* Few persons are aware that every New Guinea bird, beast, or insect we are acquainted with, has been obtained in the northern peninsulas of that country, which, as I before remarked, seem just about to be converted into islands; while the true island itself, a vast tract of forest and mountain, 800 miles long and 500 wide, is absolutely and entirely unknown. The whole of the other islands in this region which have been visited by any naturalist will not make up a tenth of this vast area; and as the mere outskirts of this unexplored land have yielded a number of remarkable genera and hosts of species which do not extend to the surrounding islands, we may be sure that the remarkable concentration of peculiar forms of Parrots which the Loriine region exhibits, even with our present imperfect knowledge of it, is very far below the reality. I believe, therefore, that we have every reason to consider New Guinea as being that still existing portion of what was once the great tropical Pacific continent to which I have alluded; and in the crimson Lories, the black *Microglossum*, the Birds of Paradise, and the great Crowned Pigeons, we have but a remnant and a sample of the strange and beautiful forms of life that once inhabited it, and many of which may still remain to be discovered in the untrodden Papuan forests.

The genus *Trichoglossus* ranges over the whole of Australia and nearly the whole of the Austro-Malayan Islands, the most remarkable exception being the northern Moluccas (Gilolo, Batchian, and Morty), which do not seem to possess it. The small species in which the sexes differ, and which are best placed with *Charmosyna*, are, however, found in these islands. Celebes, Sumatra, Timor, and Aru have each species of *Trichoglossus* peculiar to them.

The *Platycerci* are but poorly represented in the Austro-Malayan Islands, and seem to be hardly at home in the damp tropical forests. They have the same range as the Lories, but extend also to the Sulla Islands; and a species of an Australian form inhabits Timor.

We now come to the Cockatoos, another most

characteristic Australian form which ranges over the whole Australian region, except the Pacific Islands, marking out the limits of that region in the Malay archipelago by reaching Celebes and Lombock, and sending one species into the Philippine Islands, which are considered to belong to the Indian region. We must first remark that the genus *Cacatua* has a wider range in the Australian region than any other, occupying every island in the Australian and Austro-Malayan subregions, and always existing in considerable abundance. This indicates a dominant group, which has great capacities for increase and self-preservation and great powers of diffusion. It is therefore not wonderful that one species should be found to have penetrated beyond its true home. That this was due to greater facilities for emigration at a comparatively recent epoch in the existence of the genus, is indicated by the fact that, whereas throughout the rest of the archipelago the species of *Cacatua* are much restricted, each island or small group of islands possessing its peculiar form, in the Philippines one species ranges over the whole of that extensive region.

One of the most interesting genera of Parrots in the archipelago and in the world is undoubtedly *Prioniturus*, which exhibits the only instance in the whole order of a spatulate or racket-shaped tail like that of the Motmot; but in this case the perfectly bare and smooth shaft is produced by a natural process of growth, as in the King Bird of Paradise.

The four species of this remarkable genus are equally divided between Celebes and the Philippines, and present a most curious case of the restricted range of a well-marked group. An exactly analogous case among Mammalia is the genus *Cynopithecus*, a form of baboon completely unlike anything else in the East, and confined to the Philippines, Celebes, and the small adjacent island of Batchian, into which it was probably introduced. While these two groups of islands have thus evidently had once a closer connexion than at present, they both possess a striking individuality which separates them from the primary regions to which they respectively belong. The Philippines stand alone in the Indian region by the absence of all large carnivora and pachyderms, as well as of Apes and Monkeys, (in birds) by the absence of *Phasianidæ* (which are preeminently Indian) and by the presence of *Megapodius* (which is as preeminently Australian), by having no Trogons, *Palæornis* or *Eurylæmidæ*, and by possessing *Cacatua*, *Tanygnathus*, and *Cyclopsitta*. Just in a parallel manner is Celebes distinguished by the presence of peculiar forms of Antelope and Baboon, and by species of *Sciurus*, and, in birds, by having Woodpeckers, Hornbills, and several isolated genera of Passeres, while forms so characteristic of the Austro-Malayan islands as *Monarcha*, *Pachycephala*, *Tropidorhynchus*, and *Eos* are quite absent. Celebes and the Philippines will therefore form together a little intermediate region between those of Australia and India. The real cause of their distinctive peculiarities I believe to lie in their never having been immediately connected with these regions, though they have probably at some time been in closer proximity than at present--and, in the case of Celebes at least, to their representing the remains of some ancient land extending to the westward, at an epoch probably anterior to that at which Borneo and Sumatra were raised above the ocean.

The great islands which form the western half of the archipelago, Borneo, Sumatra, Java, and the Malay peninsula, present a most surprising poverty of Psittacine birds. Only four species are found over this immense region; and these belong to three genera, of which only one is found on both sides of the boundary-line. This fact forms one of the strongest proofs of the division of the archipelago between the Indian and Australian regions; for on the one side we have fifteen genera, of which ten are quite peculiar, on the other three genera, of which one is Indian, one Indo-Australian, and one somewhat isolated species only peculiar. The distribution of the genus *Loriculus*, which is the only one really common to the Indian and Australian regions, is very interesting. The southern island of the Philippines seems to be its metropolis, since no less than four species are found there; one inhabits Celebes, one Sulla, and one Gilolo; the rest are found in Flores, Java, Sumatra, and Malacca, Ceylon, India, China, and Manilla. The range of the genus is therefore very extensive; yet one-half of the species will be found concentrated in a limited tract, including Mindanao, N. Celebes, Sulla, and Gilolo. This district is upon the confines of the Australian and the Indian regions; and it is very interesting to remark that this, the only genus which is common to the two, is of doubtful affinities, and serves to connect the preeminently Australian *Trichoglossidæ* with the *Psittacidæ* of the rest of the world.

The classification and natural arrangement of the *Psittaci* has been the subject of much difference of opinion. For a long time they were placed as a simple family of Scansores along with Woodpeckers, Toucans, and Cuckoos, birds with which it is difficult to see that they have the remotest affinity, and to which they have no resemblance, except in the one character of the 2/2-toed feet.

The skull of a Parrot is remarkable for its large size, for the nearly complete orbits, for the broad and powerful lower mandible, for the large and complicated lingual and hyoid bones, and for the perfect articulation of the upper mandible to the cranium--peculiarities which, in their combination, separate it most widely from every other form of bird. The sternum has a characteristic form unlike that of any bird; the furcula is small and attached low down on the anterior margin of the keel, and in some genera is liable to be totally wanting in certain species. When present, however, it is of a semioval form, the two branches being connected in an unbroken curve without angle or projecting processes.

The prehensile feet of Parrots are used in a manner altogether peculiar; for though other birds may secure their food with their foot while eating, no others in the whole class use it systematically as a hand to grasp and convey food to the mouth. We may fairly say that they are the only birds that have hands and use them as such; and this will serve to confirm the superiority which their large brain and highly organized cranium confers upon them. The presence of a crop, their uniformly fruit-eating habits, their wide distribution, their numerous modifications of form, and their utter dissimilarity to all other birds, added to the differences already pointed out in structure and habits, induce me to adopt without any hesitation the views of Bonaparte and Blyth, and to consider the Parrots as one of the primary divisions or orders in the class of birds.

In dividing this order into families I follow generally Bonaparte and Blainville, with a few modifications for simplicity. The great central mass of the order are the *Psittacidæ* or true Parrots, comprising all the American and more than half the Old-World species. These must be divided into several subfamilies, the *Palæornithinæ*, the *Psittacinæ*, and the *Eclectinæ*, containing the Indian and Malayan species. The next family, the *Platycercidæ* (the Broadtails and Ground-Parrots), are somewhat allied to the last group through the *Palæornithinæ*. They have different habits from most other

Parrots, being often terrestrial and seed-eaters; their whole structure is weak, their flight slow and Cuckoo-like; the keel of the sternum is lower and more rounded anteriorly than in the other families; the pelvis is short, broad, and flat; the skull is small; the bill short; the lower mandible broad and swollen; the legs rather long and slender; and the plumage lax and abundant.

The *Plyctolophidæ*, or Cockatoos, are distinguished by their powerful bills, crested heads, heavy forms, and lax powdery plumage. They have a general resemblance to the last family and also to the true *Psittacidæ*. The *Trichoglossidæ* are the best-marked and most specialized group of all. The whole head, as well as the bill, is elongated and compressed; the wings long and powerful; the feet strongly grasping; and the tongue always furnished with brush-like papillæ. They are connected with the *Psittacidæ* by means of *Loriculus*, which agrees with them in general structure, but has the ordinary smooth tongue. In order to bring these families into a natural sequence, I arrange them in the following order:--1. *Plyctolophidæ*; 2. *Platycercidæ*; 3. *Psittacidæ*; 4. *Trichoglossidæ*. The fifth family, *Strigopidæ*, containing the New Zealand Owl Parrots, seems allied to the *Platycercidæ*, and should follow them in a general arrangement of the order.

I may here remark that the limits which I place to the Malayan subregion, as distinguished from the Pacific Islands--namely, to include the Solomon Islands, while the New

Hebrides and New Caledonia begin the Pacific subregion--is well established by the *Psittaci*; since both the subfamily *Loriinæ* and the family *Plyctolophidæ* reach this point only, as well as the truly Malayan genus *Geoffroyus*.

I have endeavoured to make the following list of the Malayan *Psittaci* as complete and accurate as possible. The localities have been determined from personal observation and inquiry,[2] as those usually given are very erroneous, owing to so many of the species being domesticated and carried to every part of the archipelago. Several species, which appear to have been founded on immature birds or accidental variations, are sunk altogether, as well as some which seem to have been described from made-up specimens. A few remarks on the habits of the species observed by myself are also given, and a table showing the geographical range of each species is added.

LIST OF THE MALAYAN SPECIES OF PARROTS.

Fam. I. PLYCTOLOPHIDÆ.

1. CACATUA.

1. CACATUA PHILIPPINARUM.

Psittacus philippinarum, Gm. Syst. Nat. i. ; Pl. Enl. 191.

Hab. Philippine Islands.

2. CACATUA MOLUCCENSIS.

Psittacus moluccensis, Gm. Syst. Nat. i. ; Pl. Enl. 498.

Kakadoe rubro-cristatus, Bourj. Perr. t. 78.

Plyctolophus rosaceus, Lear, Parr. pl. 2.

Hab. Ceram and Amboyna (*A. R. W.*).

Remarks.--This fine species is abundant in Ceram, but much less plentiful in Amboyna, and it is not known to extend beyond these islands. The birds are taken from the nest in holes of trees, and are easily domesticated. In Ceram they attack the young cocoa-nuts, gnawing through the tough outer covering to get at the pulp and water inside.

3. CACATUA CRISTATA.

Psittacus cristatus, Linn. Syst. Nat. i. ; Pl. Enl. 263; Bourj. Perr. t. 82.

Hab. Gilolo, Batchian, and Ternate (*A. R. W.*).

4. CACATUA CRISTATELLA, n. s.

Simillima C. cristatæ, *sed multo minor.*

Exactly like *C. cristata* in colour, but very much smaller in all its dimensions. It inhabits a limited district in the northern peninsula of Gilolo. The true *C. cristata* inhabits the other parts of Gilolo; while the specimens from Batchian and Ternate are smaller, but still seem referable to the old species. The following are the comparative dimensions of four specimens in my collection:--

	Total length. in.	Wing. in.	Tail. in.	Bill. in.
C. cristata (Gilolo)	18 1/2	12 1/4	7 1/2	1 7/8
C. cristata (Batchian)	17 1/2	10 3/4	6 3/4	1 5/8
C. cristata (Ternate), juv.?	17	10 3/4	6 1/2	1 1/2
C. cristata (Gilolo, Kao)	15	10	5 1/2	1 1/2

The iris is *red* in this species, whereas in *C. cristata* it is dark olive.

Hab. Gilolo (Kao) (*A. R. W.*).

5. CACATUA OPHTHALMICA.

Cacatua ophthalmica, Sclat. Proc. Zool. Soc. 1864, .

C. ducorpsii, Sclater, P. Z. S. 1862, , pl. xiv. (err.).

Hab. Solomon Islands? (Perhaps some adjacent island.)

6. CACATUA DUCORPSII.

Cacatua ducorpsii, H. & J. Voy. au Pôle Sud, Zool. i. , t. 26. f. l; Sclater, P. Z. S. 1864, , Pl. XVII.

Hab. Solomon Islands.

Remark.--Both these species are now living in the Gardens of the Society.

7. CACATUA TRITON.

Plyctolophus triton, Temm. Consp. Gen. Ind. Arch. iii. .

Hab. New Guinea (Goram, introduced) (*A. R. W.*).

Remarks.--Numbers of this species are brought from the mainland of New Guinea by the trading inhabitants of the little island of Goram, where escaped birds have bred, and are now very common.

8. CACATUA MACROLOPHA.

Plyctolophus macrolophus, Bernst. Cab. Journ. 1861, .

Hab. Aru Islands, Mysol, Waigiou, and Salwatty (*A. R. W.*).

Remarks.--This has been confounded with the last, and with Temminck's *C. æquatorialis.* It is intermediate in size between the two.

9. CACATUA ÆQUATORIALIS.

Plyctolophus æquatorialis, Temm. Consp. Gen. Ind. Arch. iii.

Hab. Celebes, Flores, and Lombock (*A. R. W.*).

Remarks.--This is of the same size as *C. sulphurea,* but has a much larger bill.

10. CACATUA SULPHUREA.

Psittacus sulphureus, Gm. Syst. Nat. i. ; Pl. Enl. 14; Lear. Parr. pl. 4.

Hab. Timor (*A. R. W.*).

11. CACATUA CITRINO-CRISTATA.

Plyctolophus citrino-cristatus, Fr. Proc. Zool. Soc. 1844, .

Hab. Timor-laut.

Remark.--This bird is often brought alive to Macassar in praus from the Timor-laut and Tenimber islands, to which it seems quite restricted, all the surrounding islands possessing their distinct species.

2. MICROGLOSSUM.

12. MICROGLOSSUM GOLIATH.

Psittacus goliath, Kuhl, Consp. Psitt. p, 94; Levaill. Perr. t. 12, 13.

Hab. New Guinea, Waigiou, and Mysol (*A. R. W.*).

Remarks.--This extraordinary bird is of a weak structure and feeble flight, all its muscular power seeming to be concentrated in the head and bill. It is solitary, and frequents the lower parts of the forest, and utters a plaintive whistling note, very different from the harsh scream of the true Cockatoos. The bright red colour of the face, and the long crest, which it suddenly throws up when alarmed or excited, as well as the threatening aspect of the enormous bill, may serve to frighten away birds of prey, to whose attacks its solitary habits and general weakness of structure would seem to render it especially liable. This species has probably the most powerful beak of any bird in the world, and it is the only creature that can break open the extremely hard and solid nuts of the genus *Canarium*, species of which abound in the countries it inhabits.

13. MICROGLOSSUM ATERRIMUM.

Psittacus aterrimus, Gm.; Kuhl, Consp. Psitt. p, 91 .

Ara alecto, Temm. Faun. Jap. .

P. griseus, Bechst.; Less. Perr. t. 11.

Hab. Aru Islands (*A. R. W.*); N. Australia (B.M.).

Remarks.--The two species of *Microglossum* are hardly distinguishable. The *P. aterrimus* of Gmelin seems to refer to the smaller form which inhabits the Aru Islands and N. Australia, and was therefore probably the earliest known.

3. NASITERNA.

14. NASITERNA PYGMÆA.

Psittacus pygmæus, Q. & G. Voy. de l'Astrol. t. 21. f. 1, 2; Wagl. Mon. .

Hab. New Guinea, Mysol (*A. R. W.*).

Remarks.--This curious little bird is most difficult to obtain, owing to its very small size and green colour, rendering it almost invisible among the foliage. According to the observation of my collector, Mr. Allen, it makes a hole in arboreal white ants' nests, in which it lays its eggs, like some of the small *Psittaculæ* of South America. Its spined tail would indicate some peculiar habits, of which we have as yet no account. It is a very abnormal form, and is placed among the Cockatoos with considerable doubt.

Fam. II. PLATYCERCIDÆ.

4. PLATYCERCUS.

15. PLATYCERCUS VULNERATUS.

Platycercus vulneratus, Wagl. Mon. Psitt. .

Psitt. erythropterus, Q. & G. Voy. de l'Astrol. t. 27.

Aprosmictus vulneratus, Bp.

Hab. Timor (*A. R. W.*).

16. PLATYCERCUS AMBOINENSIS.

Psittacus amboinensis, Bodd.; Briss. Orn. iv. t. 28. f. 2; Linn. Syst. Nat. i. ; Pl. Enl. 240.

Hab. Amboina, Ceram, and Bouru (*A. R. W.*).

Remarks.--This is a rather scarce bird. It has been confounded with the next species, which inhabits a different area, and is much smaller, as well as differently marked on the bill and tail. Both species eat green plantains, and are of much less sociable and tameable dispositions than most Parrots.

17. PLATYCERCUS DORSALIS.

Psittacus dorsalis, Q. & G. Voy. de l'Astrol. t. 21. f. 1.

Hab. New Guinea, Waigiou (*A. R. W.*), and Sulla Islands (var.) (*Allen*).

18. PLATYCERCUS HYPOPHONIUS.

Platycercus hypophonius, Müll. & Schleg. Verh. Nat. Gesch. .
Hab. Gilolo (*A. R. W.*).

Fam. III. PSITTACIDÆ.

5. PALÆORNIS.

19. PALÆORNIS LONGICAUDUS.
Psittacus longicaudus, Bodd. Pl. Enl. 887.
Psitt. malaccensis, Gm.; Vigors, Bp.
Hab. Malacca, Sumatra, and Borneo (*A. R. W.*).

20. PALÆORNIS JAVANICUS.
Psittacus javanicus, Osb.
Ps. osbeckii, Lath.
Palæornis barbatus, Lev. Perr. t. 31.
Hab. Java and Borneo (? *borneus*, Wagl., immature) (*A. R. W.*).

21. PALÆORNIS CANICEPS.
Palæornis caniceps, Blyth, Journ. As. Soc. Beng. xv. p, 51, 368.
Hab. Nicobar Islands (? Penang).
Remarks.--The specimen obtained by Dr. Cantor in Penang was probably brought from the Nicobar Islands, which I consider the western limit of the Malayan region.

6. PSITTINUS.

22. PSITTINUS INCERTUS.

Psittacus incertus, Shaw, Nat. Miscell. pl. 769.
P. malaccensis, Lath.; Wagl. Mon. ; Sw. Zool. Ill. pl. 154.
Hab. Malay Peninsula, Singapore, Sumatra, and Borneo (*A. R. W.*).

Remarks.--This species is the most abundant of the Parrots of the western Malay countries. The genus has undoubted affinities to *Palæornis*.

7. GEOFFROYUS.

23. GEOFFROYUS PERSONATUS.

Psittacus personatus, Shaw, Gen. Zool. viii.; Lev. Perr. t. 112, 113.
P. geoffroyanus, Vieill.
P. bataviensis, Wagl.
P. aruensis, G. R. Gray, Proc. Zool. Soc. 1858, (var.).
P. jukesii, G. R. Gray, List of Psitt. in B. M. (var.).
P. capistratus, G. R. Gray, Proc. Zool. Soc. 1858, (var.).
P. fuscicapillus, Vieill. N. Dict. d'Hist. Nat. xxv. (); Wagl. Mon.
P. spadiceocephalus, Kuhl, Consp. Psitt. .
Hab. Bouru, Ceram, Amboyna, Goram, Ké Islands (*capistratus*), Aru Islands (*aruensis*), Timor (*jukesii*), and Flores (*jukesii*) (*A. R. W.*).

Remarks.--This species varies in size, and I can find no

permanent differences by which any of the above forms can be separated.

24. GEOFFROYUS CYANICOLLIS.

Psittacus cyanicollis, Müll. & Schleg. Verh. Nat. Gesch. Ethn. p, 182.

Hab. Gilolo and Batchian (*A. R. W.*).

Remarks.--A very distinct and beautiful bird. The species of this genus are seldom domesticated. I do not remember ever having seen one in confinement.

25. GEOFFROYUS PUCHERANI.

Geoffroyus pucherani, Bp.; Souancé, Rev. et Mag. de Zool. 1856, .

Pionus fuscicapillus (Wagl.), Homb. et Jacq. Voy. au Pôle Sud, t. 25. f. 3.

Hab. New Guinea, Waigiou, and Mysol (*A. R. W.*).

26. GEOFFROYUS HETEROCLITUS.

Pionus heteroclitus, H. & J. Voy. au Pôle Sud, Zool. iii.. t. 25*. f. 1.

P. cyaniceps, Puch. Voy. au Pôle Sud, t. 25*. f. 2.

Hab. Solomon Islands.

8. PRIONITURUS.

27. PRIONITURUS FLAVICANS.

Prioniturus flavicans, Cassin, Proc. Ac. Phil. vi. **[♀];** Sclat. Proc. Zool. Soc. 1860, **[♂].**

Hab. North Celebes (Tondano) (*A. R. W.*).

Remarks.--Eats plantains about the villages; flies after dark.

28. PRIONITURUS SETARIUS.

Psittacus setarius, Temm. Pl. Col. 15.

Prioniturus platurus, Wagl. Mon. .

Psitt. spatuliger, mas, Bourj. Perr. t. 53.

Hab. Celebes (Menado and Macassar) (*A. R. W.*). *Remarks.*--Rather abundant near the town of Macassar in October. In both species the females have the spatulate tail as much developed as that of the males.

29. PRIONITURUS DISCURUS.

Prioniturus discurus, Vieill. Gal. des Ois. i., pl. 36; Wagl. Mon..

Psittacus spatuliger **[♀]**, Bourj. Perr. t. 53 *a*.

Hab. Philippine Islands (Mindanao).

30. PRIONITURUS ____.

Prioniturus spatuliger (Bourj.), G. R. G. List of Parrots in B. M. ; Sclater, Proc. Zool. Soc. 1860, .

Hab. Philippine Islands. (Probably Mindanao, the island nearest Celebes.)

9. CYCLOPSITTA.

31. CYCLOPSITTA DIOPHTHALMA.

Cyclopsitta diophthalma, H. & J. Voy. au Pôle Sud, t. 25 *bis*, f. 4, 5.

Hab. Aru Islands and Mysol (var.) (*A. R. W.*).

Remarks.--This beautiful little bird was shot while feeding on the fruit of a *ficus*, close to the trading-town of Dobbo. The specimens from Mysol have the blue spot before the eye larger, and that on the cheeks rather brighter, but less extended. What is probably an immature bird has the cheeks of the female, with the forehead of the male. More examples are wanted to determine whether this is a distinct species.

32. CYCLOPSITTA DESMARESTI.

Psittacula desmaresti, Garn. Voy. de la Coquille, Zool. t. 35; Bourj. Perr. t. 85.

Hab. New Guinea (Dorey Harbour) (*A. R. W.*).

33. CYCLOPSITTA BLYTHII, n. s.

Similis C. desmaresti, *sed capite colloque aurantiacis sine macula suboculari cærulea.*

Green; head above deep orange, more intense on the forehead; cheeks and throat pale orange; breast with a band of blue, succeeded by one of brownish orange, as in *C. desmaresti*; sides of the breast blue; under wing-coverts blue-green; belly yellowish green: bill black; feet greenish olive.

Total length 8 inches; wings 4 1/4.

Hab. Mysol (*A. R. W.*).

Remarks.--I have named this bird after Mr. Edward Blyth, who first called my attention to its distinctness from the allied species. In the British Museum Collection there is a specimen from Salwatty which I had overlooked, but which possesses a small blue cheek-spot, although in other respects it resembles this bird. Six specimens from Mysol were all exactly alike; and more than twenty specimens of *P. desmaresti*, of both sexes, collected by myself in New Guinea were equally constant. The Salwatty specimen is therefore very interesting, as showing one of the links by which these now very distinct species have been formerly connected together.

34. CYCLOPSITTA LOXIA.

Psittacus loxia, Cuv.

Psittacula loxia, Bourj. Perr. t. 94.

Hab. Philippine Islands.

35. CYCLOPSITTA LUNULATA.

Psittacus lunulatus, Scop. Sonn. Voy. t. 39.

P. torquata, Gm.; Wagl. Mon. ; Lear, Parr, pl. 39 .

Hab. Philippine Islands (Manilla).

36. CYCLOPSITTA LEUCOPHTHALMA.

Psittacus leucophthalmus, Scop. Sonn. Voy. t. 38.

P. simplex, Kuhl, Consp. p, 66.

Hab. Philippines (Luzon).

Remarks.--The genus *Cyclopsitta* is curiously divided between New Guinea and the Philippines, and seems to have its nearest external allies in the *Agapornis* of Africa.

10. TANYGNATHUS.

37. TANYGNATHUS LUCIONENSIS.

Psittacus lucionensis, Linn.; Briss. Orn. iv. t. 22. f. 2.

P. marginatus, Gm.; Wagl. Mon. .

Hab. Philippine Islands (Manilla).

38. TANYGNATHUS MEGALORHYNCHUS.

Psittacus megalorhynchus, Bodd. P. Enl. 713.

P. macrorhynchus, Gm.; Wagl. .

P. nasutus, Lath. Ind. Orn. .

Hab. Batchian, Makian, Gilolo, Mysol, Waigiou, Sanguir Islands, and New Guinea (*A. R. W.*).

39. TANYGNATHUS AFFINIS.

Tanygnathus affinis, Wallace, Proc. Zool. Soc. 1863, .

Hab. Bouru, Amboyna, and Ceram (*A. R. W.*).

40. TANYGNATHUS MÜLLERI.

Psittacus mülleri, Müll. & Schleg.Verh. Nat. Gesch. Ethn. pp.108, 182.

Hab. Celebes (Macassar and Menado) (*A. R. W.*).

41. TANYGNATHUS ALBIROSTRIS.

Psittacus sumatranus, Raffl. Linn. Trans, xiii. 281; Wagl. Mon..

Tanygnathus albirostris, Wall. Proc. Zool. Soc. 1862, .

Hab. Celebes and Sulla Islands (*A. R. W.*).

11. ECLECTUS.

42. ECLECTUS LINNÆI.

Eclectus linnæi, Wagl. Mon. , t. 22.

Hab. New Guinea, Mysol, Waigiou, and Aru Islands (*A. R. W.*).

43. ECLECTUS GRANDIS.

Psittacus grandis, Gm. S. N. i. ; Pl. Enl. 683; Wagl. Mon. .

P. ceylonensis, Bodd.

Hab. Gilolo and Batchian (*A. R. W.*).

44. ECLECTUS CARDINALIS.

Psittacus cardinalis, Bodd. Pl. Enl. 518.

P. puniceus, Gm. S. N. i. ; Wagl. Mon. ; Wall. P. Z. S. 1863, .

Hab. Bouru, Amboyna, and Ceram (*A. R. W.*).

Remark.--Buffon's figure is bad; but his description combined with it is recognizable, and evidently applies to this species. He gives Amboyna as its locality.

45. ECLECTUS CORNELIA.

Eclectus cornelia, Bp. Proc. Zool. Soc. 1849, pl. 11.

Hab. Unknown. (Probably either Ceram-laut or Jobie Islands.)

46. ECLECTUS STAVORINI.

Psittacus stavorini, Less. Voy. Coq. Zool. .

Hab. Unknown. (Perhaps Jobie Islands or N. Guinea.)

47. ECLECTUS POLYCHLOROS.

Psittacus polychloros, Scop.

P. magnus, Gm.; Wagl. Mon. ; Pl. Enl. 514.

Hab. New Guinea, Mysol, Waigiou, Aru Islands (var.), Gilolo, and Batchian (*A. R. W.*).

48. ECLECTUS INTERMEDIUS.

Psittacodis intermedius, Bp. Consp. Gen. Av. .

Hab. Ceram, Amboyna, and Bouru (*A. R. W.*).

Remarks.--This is only a smaller and rather less brightly coloured form of the last species.

49. ECLECTUS WESTERMANNI.

Psittacodis westermanni, Bp. Consp. Gen. Av. ; Proc. Zool. Soc. 1857, , pl. 127.

Hab. Unknown. (Probably New Guinea or Jobie Islands.)

Note.--The red and the green-coloured species of this genus are so alike in structure and habits that it is useless to separate them by adopting the genus *Psittacodis* for the latter. They are dull and heavy birds, frequenting low trees and the neighbourhood of villages, devouring plantains and papaw-fruits, and even descending to feed upon the red peppers, which they sometimes gorge so that they may be taken by hand.

12. DASYPTILUS.

50. DASYPTILUS PEQUETII.

Psittacus pequetii, Less. Bull. Univers. 1831, .

Hab. New Guinea and Salwatty.

Remarks.--A living specimen of this rare and curious bird was seen at Salwatty by my assistant, Mr. Allen, in the possession of a Bugis trader, who would not part with it. The natives knew it, but said it was very rare. Its affinities are doubtful, and it may perhaps belong to the next family.

13. LORICULUS.

51. LORICULUS GALGULUS.

Psittacus galgulus, Linn.; Wagl. Mon. ; Pl. Enl. 190.

Hab. Malacca, Sumatra, and Borneo (*A. R. W.*).

Remarks.--At certain seasons this exquisite little bird is very plentiful, and is captured by the natives, soon becoming tame. It sleeps with its head downwards, suspended by one leg.

52. LORICULUS STIGMATUS.

Psittacus stigmatus, Müll. & Schleg. Verh. Nat. Gesh. Ethn. .

Hab. Celebes (Macassar and Menado) (*A. R. W.*).

53. LORICULUS SCLATERI.

Loriculus sclateri, Wall. Proc. Zool. Soc. 1862, , pl. 38.

Hab. Sulla Islands (*A. R. W.*).

Remarks.--This, one of the finest species of the genus, is only known from some small islands on the east side of Celebes.

54. LORICULUS AMABILIS.

Loriculus amabilis, Wall. Ibis, 1862, .

Hab. Gilolo (*A. R. W.*).

55. LORICULUS PUSILLUS.

Loriculus pusillus, G. R. Gray, Brit. Mus. Cat. Psitt. .

Psittacus vernalis, Kuhl; Sw. Zool. Ill. 2nd ser. pl. 1.
Hab. Java (*A. R. W.*).
Remarks.--I found this bird myself in Java. Its simple colouring has led to its being taken for other species in an immature state. The sexes are alike.

56. LORICULUS FLOSCULUS.
Loriculus flosculus, Wall. Proc. Zool. Soc. 1863, .
Hab. Flores (*A. R. W.*).

57. LORICULUS MELANOPTERUS.
Psittacus melanopterus, Scop. Sonn. Voy. t. 40.
P. minor, Gm.
Hab. Philippines (Mindanao).

58. LORICULUS APICALIS.
Loriculus apicalis, Souancé, Rev. et Mag. de Zool. 1856, .
Hab. Philippines (Mindanao).

59. LORICULUS BONAPARTEI.
Loriculus bonapartei, Souancé, Rev. de Zool. 1856, .
Psittacus minor, Wagl. Mon. (nec Gmel.).
Hab. Sooloo Islands.

60. LORICULUS CULACISSI.
Psittacus culacissi, Vieill. Enc. Méth. ; Wagl. Mon. .

P. rubrifrons, Vig.; Lear, Parr. pl. 41.
P. philippensis, Steph. Gen. Zool. xiv. .
Hab. Philippine Islands.

61. LORICULUS REGULUS.
Loriculus regulus, Souancé, Rev. et Mag. de Zool. 1856, .
Hab. Philippines (Mindanao).

Fam. IV. TRICHOGLOSSIDÆ.

14. LORIUS.

62. LORIUS DOMICELLA.
Psittacus domicella, Linn.; Pl. Enl. 119; Wagl. Mon. .
Hab. Ceram and Amboyna (*A. R. W.*).
Remarks.--The *Psittacus raja* of Shaw, Vieillot, &c., is probably a domesticated variation of this species, induced by peculiar food.

63. LORIUS LORY.
Psittacus lory, Linn.; Pl. Enl. 168.
Domicella lory, Wagl. Mon. .
Lorius tricolor, Steph.
Hab. New Guinea, Waigiou, and Mysol (*A. R. W.*).

64. LORIUS CYANAUCHEN.

Lorius cyanauchen, Müll. & Schleg. Verh. Nat. Gesch. Ethn. .
L. superbus, Fraser, Proc. Zool. Soc. 1845, .
Hab. Myfor and Jobie islands (New Guinea).
Remarks.--While I was staying at Dorey Harbour a native vessel arrived from Myfor Island, bringing a few of these fine birds; and I was assured they were also found in Jobie.

65. LORIUS GARRULUS.
Psittacus garrulus, Linn.
Domicella garrula, Wagl. Mon. ; Pl. Enl. 216.
Hab. Gilolo and Batchian (*A. R. W.*).
Remarks.--The yellow dorsal patch varies much in most Gilolo birds, being very small; while in Batchian specimens it is always very large and conspicuous.

66. LORIUS CHLOROCERCUS.
Lorius chlorocercus, Gould, Proc. Zool. Soc. 1856, .
Hab. Solomon Islands.

67. LORIUS HYPŒNOCHROUS.
Lorius hypoinochrous, G. R. Gray, List of Psittacidæ in Brit. Mus. .
Hab. Louisiade archipelago.
Note.--Lorius cæruleatus and *L. cyanurus* are probably artificial birds.

15. CHALCOPSITTA.

68. CHALCOPSITTA ATRA.

Psittacus ater, Scop.

P. novæ-guineæ, Gm.; Wagl. Mon. ; Lev. Perr. t. 49.

Chalcopsitta bernsteini, Rosenb. Cab. Jour. 1858, .

Hab. New Guinea and Mysol (*A. R. W.*).

69. CHALCOPSITTA SCINTILLATA.

Psittacus scintillatus, Temm. Pl. Col. 569 (juv.); Bourj. Perr. t. 51.

Chalcopsitta rubrifrons, G. R. Gray, Proc. Zool. Soc. 1858, , pl. 135.

Hab. New Guinea (*Temm.*) and Aru Islands (*A. R. W.*).

Remarks.--This beautiful bird is very rare. Both this and the preceding species are easily domesticated, and are universal favourites from their good temper and docility.

70. CHALCOPSITTA RUBIGINOSA.

Chalcopsitta rubiginosa, Bp. Consp. Av. ; Proc. Zool. Soc. 1850, pl. 16.

Hab. Unknown.

16. EOS.

71. EOS INDICA.

Psittacus indicus, Gm.; Pl. Enl. 143; Lev. Perr. t. 53.
P. coccineus, Lath.
Hab. Siau and Sanguir.
Remarks.--This beautiful bird is brought by native traders to Menado and Ternate from these islands, which are its only known habitat; it must therefore be considered as belonging to the Moluccas, although its habitat is situated between Celebes and the Philippines. *Psittacus variegatus*, Gm., is probably a made-up bird.

72. EOS CYANOGENIA.
Eos cyanogenia, Bp. Consp. Av. ; Proc. Zool. Soc. 1850, pl. 14.
Hab. Myfor and Jobie islands (N. of New Guinea).
Remarks.--Living specimens of this bird were brought to Dorey along with *Lorius cyanauchen*.

73. EOS CYANOSTRIATA.
Eos cyanostriata, G. R. Gray, Gen. of Birds, ii. pl. 103.
Lorius borneus, Less.
Hab. Tenimber Islands and Timor-laut.
Remarks.--This species is often brought alive to Macassar in the Bugis praus, which go to the Tenimber Islands for the tripang fishery.

74. EOS RUBRA.

Psittacus ruber, Gm. (*P. borneus*, L.); Pl. Enl. 519; Lev. Perr. t. 44.

P. moluccensis, Lath.

P. cæruleatus, Shaw; Lev. Perr. t. 93.

Hab. Ceram, Goram, Amboyna, and Bouru (*A. R. W.*).

75. EOS SEMILARVATA.

Eos semilarvata, Bp. Consp. Av. ; Proc. Zool. Soc. 1850, pl. 15.

Hab. Unknown. (Not Ceram, probably Timor-laut.)

76. EOS SQUAMATA.

? *Psittacus squamatus*, Bodd. Pl. Enl. 684.

? *P. guebiensis*, Scop.

Hab. Waigiou (*A. R. W.*).

Remarks.--This seems distinct from the next, having constantly much narrower bands on nape and less purple beneath. Its identification as above is doubtful.

77. EOS RICINIATA.

Psittacus riciniatus, Bechst.; Lev. Perr. t. 54.

P. cochinsinensis, Lath.; ? Wagl. Mon. .

Hab. Batchian and Gilolo (*A. R. W.*).

78. EOS FUSCATA.

Eos fuscata, Blyth, J. A. S. Bengal, 1858, . *Eos torrida*, G. R.

Gray, List of Psittacidæ in B. M. .

Hab. New Guinea (Dorey) (*A. R. W.*).

Remarks.--The yellow variety of this bird is well described by Mr. Blyth, and his dimension ("wing 6 in.") is correct. The two sexes of both red and yellow varieties were obtained from one flock, which visited Dorey for a few days only during my residence there.

79. EOS UNICOLOR.

Psittacus unicolor, Shaw; Lev. Perr. t. 125.

Lorius cardinalis, H, & J. Voy. au Pôle Sud, Zool. iii. , t. 24*. f. 2.

Hab. Solomon Islands.

17. TRICHOGLOSSUS.

80. TRICHOGLOSSUS HÆMATODUS.

Psittacus hæmatodus, Linn.; Lev. Perr. t. 47; Wagl. Mon. .

P. capistratus, Bechst.; Kuhl, Consp. Psitt.

Hab. Timor (*A. R. W.*).

81. TRICHOGLOSSUS FORSTENI.

Trichoglossus forsteni, Bp. Consp. Av. .

Hab. Sumbawa (*Bonaparte*).

82. TRICHOGLOSSUS CYANOGRAMMUS.

Trichoglossus cyanogrammus, Wagl. Mon. Psitt. .

Psittacus hæmatotus, Gm. S. N. i. ; Pl. Enl. 61.

T. nigrogularis, G. R. Gray, Proc. Zool. Soc. 1858, .

Hab. New Guinea, Waigiou, Mysol, Matabello Islands, Goram, Ceram, Amboyna, Bouru, and Aru Islands (*A. R. W.*).

Remarks.--This, like most species of wide ranges, varies irregularly, and I can find no character distinguishing the specimens from Aru.

83. TRICHOGLOSSUS COCCINEIFRONS.

Trichoglossus coccineifrons, G. R. Gray, P. Z. S. 1858, .

? *Trichoglossus immarginatus*, Blyth, J. A. S. Bengal, 1858, .

Hab. Aru Islands (*A. R. W.*).

Remarks.--Mr. Blyth's bird is probably the same as this, but immature, as he does not mention the red on the head. In other details of colouring it seems to agree, especially in the red colour of the under surface of the wings.

84. TRICHOGLOSSUS ORNATUS.

Psittacus ornatus, Linn.

Eos ornata, Wagl. Mon. .

Lorius ornatus, Steph.

Australasia malaisiæ, Less. Tr. d'Orn. .

Hab. Celebes (*A. R. W.*).

Remarks.--This beautiful species abounds all over Celebes, to which island it seems restricted.

85. TRICHOGLOSSUS FLAVOVIRIDIS.

Trichoglossus flavoviridis, Wallace, Proc. Zool. Soc. 1862, .

Hab. Sulla Islands and ? Celebes (Menado) (*A. R. W.*).

Remarks.--I obtained a single specimen at Menado, which was unfortunately eaten by rats, but seemed to be the same as this species.

86. TRICHOGLOSSUS EUTELES.

Psittacus euteles, Temm. Pl. Col. 568.

Trichoglossus ochrocephalus, Blyth, J. A. S. Bengal, 1858, .

Hab. Timor and Flores (*A. R. W.*).

Remarks.--This bird was very abundant in Timor, frequenting the blossoms of the *Eucalypti*. I see no reason for separating it generically from the other species as has been done by Bonaparte, who calls it *Psitteuteles euteles*.

87. TRICHOGLOSSUS IRIS.

Psittacus iris, Temm. Pl. Col. 567.

Hab. Timor (*A. R. W.*).

88. TRICHOGLOSSUS MASSENA.

Trichoglossus massena, Bp. Rev. et Mag. de Zool. 1854, .

Hab. Solomon Islands.

18. CHARMOSYNA.

89. CHARMOSYNA PAPUENSIS.

Psittacus papuensis, Gm. (*japonicus*, L.).

P. bontii, Lath.

Lorius papuensis, Less. Tr. d'Orn. ; Lev. Perr. t. 77.

Hab. New Guinea (*A. R. W.*).

Remarks.--I saw tail-feathers at Dorey which had been obtained on the spot by the natives.

90. CHARMOSYNA PULCHELLA.

Charmosyna pulchella, G. R. Gray, List of Psitt. in B. M. .

Hab. New Guinea (Dorey) (*A. R. W.*).

91. CHARMOSYNA PLACENTIS.

Psittacus placentis, Temm. Pl. Col. 553.

Hab. Batchian and Gilolo (*A. R. W.*).

Var. *a*. With less red on throat.

Hab. Ceram (*A. R. W.*).

Var. *b*. With scarcely any red on throat.

Hab. Aru Islands (*A. R. W.*).

92. CHARMOSYNA RUBRONOTATA.

Coriphilus rubronotatus, Wall. Proc. Zool. Soc. 1862, .

Hab. Salwatty (*A. R. W.*).

Remarks.--I have included the last two species in this genus rather than in *Trichoglossus* or *Coriphilus*, because they agree in their small size, acute red- or yellow-tipped middle tail-feathers, red or yellow feet, an probably in the sexes differing, as in *C. placentis*.

	Malacca.	Sumatra.	Borneo.	Java.	Philippines.	Celebes.	Sulla Islands.	Lombock.	Sumbawa.	Flores.	Timor.	Bouru.	Ceram.	Goram.	Timor-laut.	Batchian.	Gilolo.	Sanguir Islands.	Aru Islands.	Mysol.	Waigiou.	Jobie and Myfor.	Salwatty.	New Guinea.	Solomon Islands, and those east of New Guinea.
PLYCTOLOPHIDÆ.																									
Cacatua																									
philippinarum, *Gm.*					1																				
moluccensis, *Gm.*													1												
cristata, *L.*																1	1								
cristatella, *Wall.*																	1								
ophthalmica, *Sclat.*																									1
ducorpsii, *H. & J.*																									1
triton, *Temm.*														1							1			1	
macrolopha, *Bernst.*																			1	1	1		1		
æquatorialis, *Temm.*						1		1	1	1															
sulphurea, *Gm.*											1														
citrinocristata, *Fr.*															1										
Microglossum																									
goliath, *Temm.*																				1	1		1	1	
aterrimum, *Gm.*																			1						
Nasiterna																									
pygmæa, *Q. & G.*																				1				1	
PLATYCERCIDÆ.																									
Platycercus																									
vulneratus, *Wagl.*											1														
amboinensis, *Bodd.*												1	1												
dorsalis, *Q. & G.*							1														1			1	
hypophonius, *M. & S.*																	1								
PSITTACIDÆ.																									
Palæornis																									
longicaudus, *Bodd.*	1	1	1																						
javanicus, *Osb.*			1	1																					
caniceps, *Blyth*	1																								
Psittinus																									
incertus, *Shaw*	1	1	1																						
Geoffroyus																									
personatus, *Shaw*										1	1	1	1	1					1						
cyanicollis, *Müll.*																1	1								
pucherani, *Bp.*																				1	1			1	
heteroclitus, *H. & J.*																									1
Prioniturus																									
flavicans, *Cass.*						1																			
setarius, *Temm.*						1																			

	Malacca.	Sumatra.	Borneo.	Java.	Philippines.	Celebes.	Sulla Islands.	Lombock.	Sumbawa.	Flores.	Timor.	Bouru.	Ceram.	Goram.	Timor-laut.	Batchian.	Gilolo.	Sanguir Islands.	Aru Islands.	Mysol.	Waigiou.	Jobie and Myfor.	Salwatty.	New Guinea.	Solomon Islands, and those east of New Guinea.
PSITTACIDÆ (*continued*).																									
Prioniturus																									
discurus, *Vieill.*					1																				
spatuliger, *Bourj.*					1																				
Cyclopsitta																									
diophthalma, *H. & J.*																			1	1					
desmaresti, *Garn.*																								1	
blythii, *Wall.*																				1			1		
loxia, *Cuv.*					1																				
lunulata, *Scop.*					1																				
leucophthalma, *Scop.*					1																				
Tanygnathus																									
lucionensis, *L.*					1																				
megalorhynchus, *Bodd.*																1	1	1		1	1			1	
affinis, *Wall.*												1	1												
müller, *Schl.*						1																			
albirostris, *Wall.*						1	1																		
Eclectus																									
linnæi, *Wagl.*																			1	1	1			1	
grandis, *Gm.*																1	1								
cardinalis, *Bodd.*												1	1												
cornelia, *Bp.*																									
stavorini, *Less.*																									
polychloros, *Scop.*																1	1		1	1	1			1	
intermedius, *Bp.*												1	1												
westermanni, *Bp.*																									
Dasyptilus																									
pecquetii, *Less.*																							1	1	
Loriculus																									
galgulus, *L.*	1	1	1																						
stigmatus, *Müll.*						1																			
sclateri, *Wall.*							1																		
amabilis, *Wall.*																	1								
pusillus, *G. R. G.*				1																					
flosculus, *Wall.*										1															
melanopterus, *Scop.*					1																				
apicalis, *Souancé*					1																				
bonapartei, *Souancé*					1																				
culacissi, *Vieill.*					1																				
regulus, *Souancé*					1																				
TRICHOGLOSSIDÆ.																									
Lorius																									
domicella, *L.*													1												
lory, *L.*																				1	1		1	1	
cyanauchen, *Müll.*																						1			
garrulus, *L.*																1	1								
chlorocercus, *Gould*																									1
hypœnochrous, *G. R. G.*																								1	
Chalcopsitta																									
atra, *Scop.*																				1				1	
scintillata, *Temm.*																			1					1	
rubiginosa, *Bp.*																									

	Malacca.	Sumatra.	Borneo.	Java.	Philippines.	Celebes.	Sulla Islands.	Lombock.	Sumbawa.	Flores.	Timor.	Bouru.	Ceram.	Goram.	Timor-laut.	Batchian.	Gilolo.	Sanguir Islands.	Aru Islands.	Mysol.	Waigiou.	Jobie and Myfor.	Salwatty.	New Guinea.	Solomon Islands, and those east of New Guinea.
TRICHOGLOSSIDÆ (*cont.*).																									
Eos																									
indica, *Gm.*																		1							
cyanogenia, *Bp.*																						1			
cyanostriata, *G. R. G.*															1										
rubra, *Gm.*												1	1	1											
semilarvata, *Bp.*																									
squamata, *Bodd.*																					1				
riciniata, *Bechst.*																1	1								
fuscata, *Bl.*																								1	
unicolor, *Shaw*																									1
Trichoglossus																									
hæmatodus, *L.*											1														
forsteni, *Bp.*									1																
cyanogrammus, *Wagl.*												1	1	1					1	1	1			1	
coccineifrons, *G. R. G.*																			1						
ornatus, *L.*						1																			
flavoviridis, *Wall.*						1	1																		
euteles, *Temm.*										1	1														
iris, *Temm.*											1														
massena, *Bp.*																									1
Charmosyna																									
papuensis, *Gm.*																								1	
pulchella, *G. R. G.*																								1	
placentis, *Temm.*													1			1	1		1						
rubronotata, *Wall.*																							1		
Species in each island	3	3	4	2	12	8	4	2	2	4	6	7	10	4	2	8	11	2	10	12	11	2	6	18	6
Sp. peculiar to each island	0	0	0	1	..	5	1	0	1	1	4	0	2	0	2	0	3	1	2	0	1	2	1	4	
Genera in each island	3	3	3	2	5	5	4	2	2	4	4	6	9	..	..	..	9	..	..	10	..	..	..	14	

	Mal., Sum., Bor., & Java.	Phil.	Celebes.	Timor group.	Moluccas group.	Papuan group.
Sp. in groups of islands	5	12	10	9	23	35
Peculiar species	5	12	8	7	18	31
Genera	3	5	6	4	10	14
Peculiar genera	1	0	0	0	0	4
		1 genus. (Phil. and Celebes)				
	One genus quite peculiar.	10 genera peculiar to the Austro-Malayan region!				

18 genera in the Archipelago (13 peculiar).

Notes Appearing in the Original Work

1. The red line on the accompanying map encloses the area to which the crimson Lories are restricted.

2. N.B. The initials (*A. R. W.*) after any locality show that the species was observed there by myself.

Made in United States
Orlando, FL
04 November 2026

90218649R00032